W9-CHC-770

Seasons in a Year

SPRING

AMY CULLIFORD

A Crabtree Roots Book

CRABTREE
Publishing Company
www.crabtreebooks.com

School-to-Home Support for Caregivers and Teachers

This book helps children grow by letting them practice reading. Here are a few guiding questions to help the reader with building his or her comprehension skills. Possible answers appear here in red.

Before Reading:

- What do I think this book is about?
    - *This book is about a season called spring.*
    - *This book is about things you can see in spring.*

- What do I want to learn about this topic?
    - *I want to learn what spring looks like.*
    - *I want to learn what happens in spring.*

During Reading:

- I wonder why...
    - *I wonder why it rains so much in spring.*
    - *I wonder why bugs come out in spring.*

- What have I learned so far?
    - *I have learned what spring looks like.*
    - *I have learned that grass grows in spring.*

After Reading:

- What details did I learn about this topic?
    - *I have learned that flowers come out in spring.*
    - *I have learned that it rains a lot in spring.*

- Read the book again and look for the vocabulary words.
    - *I see the word **flowers** on page 5 and the word **rain** on page 12. The other vocabulary words are found on page 14.*

What do you see in **spring**?

4

I see white **flowers**.

I see red flowers.

I see **grass**.

I see **bugs**.

I see **rain**.

Word List

Sight Words

a
do
I
in
red
see
what
you

Words to Know

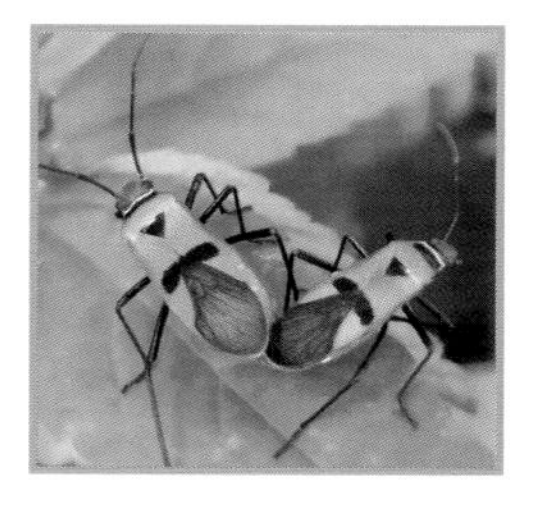

bugs

flowers

grass

rain

spring

23 Words

What do you see
in **spring**?

I see white **flowers**.

I see red flowers.

I see **grass**.

I see **bugs**.

I see **rain**.

Written by: Amy Culliford
Designed by: Rhea Wallace
Series Development: James Earley
Proofreader: Kathy Middleton
Educational Consultant: Christina Lemke M.Ed.
Photographs:
Shutterstock: mashe: cover; nnattalli: p. 1; Olya Humeniuk: p. 3, 14; natalia bulatova: p. 4; Tamotsulto: p. 14; Aeypix: p. 7; Nick Beer: p. 8-9, 14; udin fotography: p. 11, 14; Stone36: p. 13, 14

Seasons in a Year

SPRING

Library and Archives Canada Cataloguing in Publication
Title: Spring / Amy Culliford.
Names: Culliford, Amy, 1992- author.
Description: Series statement: Seasons in a year | "A Crabtree roots book".
Identifiers: Canadiana (print) 20200387057 | Canadiana (ebook) 20200387065 | ISBN 9781427134738 (hardcover) | ISBN 9781427132703 (softcover) | ISBN 9781427132741 (HTML) | ISBN 9781427133120 (read-along ebook)
Subjects: LCSH: Spring—Juvenile literature.
Classification: LCC QB637.5 .C85 2021 | DDC j508.2—dc23

Library of Congress Cataloging-in-Publication Data
Names: Culliford, Amy, 1992- author.
Title: Spring / Amy Culliford.
Description: New York : Crabtree Publishing Company, 2021. | Series: Seasons in a year : a Crabtree roots book | Audience: Ages 4-6 | Audience: Grades K-1 | Summary: "Early readers are introduced to the spring season. Simple sentences and engaging pictures bring the season of new growth alive"-- Provided by publisher.
Identifiers: LCCN 2020049779 (print) | LCCN 2020049780 (ebook) | ISBN 9781427134738 (hardcover) | ISBN 9781427132703 (paperback) | ISBN 9781427132741 (ebook) | ISBN 9781427133120 (epub)
Subjects: LCSH: Spring--Juvenile literature.
Classification: LCC QB637.5 .C85 2021 (print) | LCC QB637.5 (ebook) | DDC 508.2--dc23
LC record available at https://lccn.loc.gov/2020049779
LC ebook record available at https://lccn.loc.gov/2020049780

Crabtree Publishing Company
www.crabtreebooks.com 1-800-387-7650

Printed in Canada/012021/CPC20210118

Published in the United States
Crabtree Publishing
347 Fifth Avenue, Suite 1402-145
New York, NY, 10016

Published in Canada
Crabtree Publishing
616 Welland Ave.
St. Catharines, Ontario L2M 5V6